Drawn
to Nature

Don Conroy is a household name in Ireland – just hearing 'draw with Don' evokes nostalgic memories for many. A comforting presence on Irish television for decades, Don has inspired generations of children with the joy of drawing and creativity. A writer, a television presenter, a naturalist and a working artist, Don travels the country, passionately promoting the importance of art and conservation.

Drawn to Nature

Encounters with Irish Wildlife

DON CONROY

GILL BOOKS

Gill Books
Hume Avenue
Park West
Dublin 12
www.gillbooks.ie

Gill Books is an imprint of M.H. Gill and Co.

978 18045 8204 6

Designed by Bartek Janczak
Proofread by Emma Dunne
Printed and bound by Firmengruppe APPL, Germany
This book is typeset in 11 on 16pt, Minion Pro.

The paper used in this book comes from the wood pulp of sustainably managed forests.

A CIP catalogue record for this book is available from the British Library.

5 4 3 2 1

To my wonderful wife Gay, who
has always shown such great
support in all my creative
endeavours down through
the years.

To my daughters Sarah, Sophie
and especially Justine for her
wonderful support on this project.

To my sons Richie and David,
who helped spark off the idea.

Contents

Introduction

To experience the changing face of the landscape is as much of a joy to the mind as it is to the eye. We are privileged to live in a part of the world where we can enjoy a climate that possesses its own unique, distinctive four seasons. Each season has a treasure trove of things to see, hear, observe and marvel at. So much to enjoy, learn and wonder about. A pageant of beauty passing by for eyes willing to see, ears willing to listen. Sometimes showy, other times rather subtle. Allow your senses to awaken, look with fresh and creative eyes – there is a source of beauty ready to awaken the creative mind (which I believe we all possess).

Down through the ages, artists have been rewarded by nature with dazzling visions and deep insights, their imaginations enriched, helping them to capture the colours, forms and shapes. Whether they be writers, poets, musicians, weavers, potters, all have benefitted from their encounters with nature. I invite you to drink from a deeper well of understanding. The great Michelangelo said there was a secret shape and form in everything.

Science will explain the complex, interwoven web that is life in all its rich and varied forms and which has enriched our understanding of the natural world. Yet for me, nature seems more than a rich spectrum of flora and fauna. I believe it has the power to calm the mind, nourish the soul. Sometimes when I step into the landscape I am walking on hallowed ground; in the silence of the woods, one may get the sense of the sacred, inviting us to walk the pilgrim's way, along some sacred path in the dappled sunlight.

INTO THE WOODS

Take time to observe
wildlife — be it a fleeting
glimpse, or a chance
to really watch and
observe the activity. All
experiences are memorable
and may result in a
sketch, or even a painting,
or a poem.

Red Squirrel

Brownish-red in colour with a creamy throat, ear tufts and a bushy tail, red squirrels are usually found in coniferous woodlands.

I have had the pleasure of watching these delightful creatures in such places as Raven Wood in Co. Wexford and the beautiful Phoenix Park in Dublin.

Grey Squirrel

This species, native to North America, was introduced to Ireland in 1911 and soon spread to most of the counties in the country. Larger and more adaptable, it began to threaten our own native species, the red squirrel.

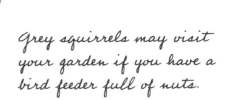

Grey squirrels may visit your garden if you have a bird feeder full of nuts.

Let us take time to reclaim the beauty of nature in our own lives.

Ancient winter woods

The woods are silent
And cloaked in shadows
Trees stand in sublime stillness
Branches ink across a starry night
Enter Diana, the moon goddess
Elegantly tracing her way
Against the ancient sky pathways
Stars twinkle in the nourishing silence.

A cool breeze quivers small twigs
Without breaking the quiet solitude
Nearby stirring in an old pine
Barely perceived
An owl stands sentinel-like on an outer branch
Shuffles its feathers
Like a shade
Takes off into the silvery light
Gliding slowly on hushed wings
As if to embrace the interior darkness of the woods.

What arcane secrets are housed there?
Archives of memories?
Time to leave.
A snap of a twig beneath my foot.
Hush! A voice seems to whisper on the breeze
The woods are dreaming.

Red Fox

I simply love to draw this very attractive creature and I admire its ability to not only survive but flourish in the country despite the odds.

I lived for a time in Celbridge in Co. Kildare. There I got to know a fox by leaving food out for it at the edge of the woods. Getting used to my scent and my presence, he would eat in front of me, occasionally brushing my leg like a cat.

One

Two

Three

Four

11

Nature reveals the common, the rare, the magnificent.

Raven

Juvenile great
spotted woodpecker

Long-Eared Owl

Despite the name, these owls do not have long ears. The feather tufts on top of the head are for communication with one another. Their flight is silent and they flatten their ear tufts when flying. They are not very good at building nests and prefer to take over an old crow or magpie nest.

Go to the woods in the late springtime and you may hear the low, muffled hoot of this owl.

One

Two

Three

Four

Long-eared owl

Woodcock

Woodcocks are rather unusual; they are waders, but they live in the woodlands, not along estuaries and shorelines as most other wading birds do. They have cryptic plumage, making them almost impossible to spot.

The first time I saw a woodcock was after encountering two hunters. One man was holding this beautiful but lifeless bird. 'It's a tasty little game bird, you need to be a good shot to bag one,' the other remarked.

Tree Creeper

Easily overlooked, this unobtrusive bird with a curved bill moves mouse-like up the tree trunk in search of hidden grubs and insects.

Some tree creepers will hollow out a winter nest in soft bark trees, such as Wellingtonia.

Pine Marten

The pine marten has a dark-brown fur coat with a reddish hue on the belly. It has a long snout with a yellowish patch underneath the chin extending to the chest, big feet with sharp claws and a long bushy tail.

I was once asked to do a radio programme on this fascinating creature. A wildlife ranger invited us to his home having informed us that he had regular visits from a family of pine martens to his wild garden. We sat in the darkened kitchen at dusk, enjoying some apple tart and tea kindly provided by his lovely wife. Staring out into the garden for a glimpse of these shy creatures, the ranger's six-year-old son asked me if I liked peanut butter. Surprised, I said I did. 'So do the pine martens,' he retorted.

We were rewarded with the amazing sight of four pine martens in the garden, scouting for the dollops of peanut butter that the ranger had placed in various spots on logs and branches. The martens enjoyed the treats; we enjoyed the wonderful views. Now I cannot eat a peanut-butter sandwich without thinking of that magical evening.

Badger

With the extinction of such predators as the wolf, lynx and bear, badgers can claim the title of the largest carnivore in Ireland. They can also boast of the most powerful bite of any animal in the country.

They are very sociable creatures and there may be as many as twenty sharing a sett, which has a series of entries and tunnels.

Giving a badger a lift home

Having just drifted off to sleep, I thought I heard a tapping on a window, faint at first, then it seemed to become louder, more like a rapping. I checked the clock; it was after midnight. There was silence … then the tapping continued, this time on the hall door. With my wife and three children asleep, I headed down the stairs to investigate. Cautiously, I opened the front door of our home in Co. Kildare.

'Ah, how are ya, sorry for troubling you at this hour. But my girl-friend told me where you lived. Well, she's not really my girlfriend, only a date. I met her at a dance in town and I offered her a lift home not realising she wasn't from the city but lived in Clane.'

At this stage I was beginning to get cold with the sharp wind whistling past me in the doorway. I could smell the drink from the man's breath. I invited him in, offering him tea and biscuits, feeling he was not fit to drive in this state.

'I saw you on *The Late Late Show*,' he said delightedly. Having supplied us both with tea, I asked why he had called to our home.

'Ah well, yes, you see I love animals. I even keep racing pigeons. Sorry, I'll cut to the chase … I know it's late and you're in your pyjamas and all. Me having to get to work early in the morning, you know. You see I was driving Angela home, that's her name, Angela … she lives in Clane, a bit of a foreigner for me, well, as I was telling you, I was driving her home when a bleeding dog ran out in front of the car. She was upset and so was I and I said I love animals. I got out and threw an old coat over it and put it in the boot. She said she didn't want her father up at that hour of the night then she thought of you, being a lover of birds and animals and on *The Late Late Show* and all.'

I told him to bring it into the house and if it survived the night I would bring it to the vet first thing in the morning. He returned with a rather smelly army-style coat and left it in the hallway.

'Look, I really appreciate this, you're a gem,' he said, shaking my hand.

'Are you sure you are fit to drive?'

'Ah sure, that cuppa tea has me right as rain. Can I give you a couple of bob for the vet?'

'You're fine,' I said. 'Safe journey home.'

He grinned. 'Next time I'll check out where the bird lives before I offer a lift home,' then he left.

I lifted back the coat and the animal shot out from the covering down the hallway to the kitchen, slipping on the kitchen tiles and knocking over a stool. I couldn't believe my eyes. It was not some farm dog but a badger. It puffed itself up and began to make snapping sounds. In a panic I threw the heavy army coat back over it then carefully carried it to our empty garage, thinking it couldn't do much damage there.

The next morning it was much calmer and I released it into the back garden where it immediately began to dig up the lawn, grubbing for worms. I fed it with some cereal and milk, which it enjoyed. Our children were delighted to see this amazing animal that had joined us for breakfast. My wife was used to people bringing injured birds, but a badger was another story. Realising there wasn't anything wrong with the badger, I decided to drive to Clane and return it to the area where it had been found. The children wanted to keep it. I explained that it was a wild creature and needed to return to its family. Having got a large cardboard box from the electrical shop, I threw the smelly coat back over the badger and both were placed inside the box, with a few small holes to let air in. I was driving a Renault 4 at the time, and I loaded up this unexpected guest into the boot and headed for Clane.

Happy in the knowledge that we were getting this feisty animal out of our garden back to its home in the wild, travelling along, the journey seemed fine and there was very little sound coming from the box. I was enjoying the view and listening to my favourite music, but as I got near the destination, the cardboard box began to suddenly rock about. The next minute I discovered that the badger had torn through the cardboard and was now on the back seat of the car. In desperation to get away, it headed for the windscreen. Luckily, I was driving on a narrow rural road with little traffic coming against me. I did my best to hold on to the steering wheel while the badger climbed all over me, trying to find a way out. A car passed on the other side of the road. I can only imagine what the driver must have thought when he looked over into my car to see a badger in the passenger seat. Finally I managed to stop near the entrance of the woods. I quickly jumped out, opening all the doors and the boot. Slowly, the badger left the car, trundled across the road and sniffed the air. It seemed to stare at me, then slipped away into the long grasses towards the woods.

It was my very first encounter with a wild badger, but not my last.

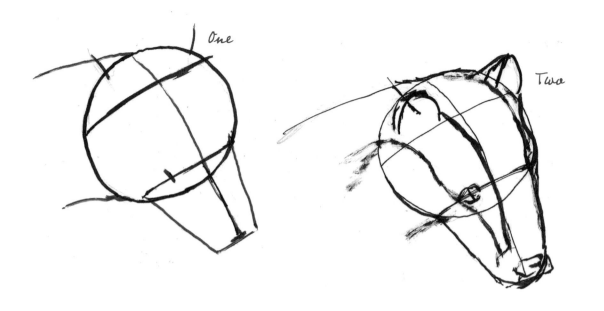

One

Two

Three

Four

Fallow Deer

When recreational parks were created, the fallow deer was ideal for grazing the long grasses. Some escaped and successfully lived in the wild. Their coats can vary from chestnut brown with white spots to light brown with spots.

Sessile Oak

The mighty oak tree is a wonderful sight to behold. While out for a walk with John Moriarty, Irish poet, philosopher and writer, we stopped to admire a fine specimen.

'Don, I've passed this tree on many occasions, and you know, I'm convinced that this tree is wiser than us, having seen what it has over the years ...'

35

Looking to nature is not just about accumulating knowledge. It may even help us reach a transcendental awakening, an exalted moment, if we allow ourselves the experience.

GOING WITH THE FLOW

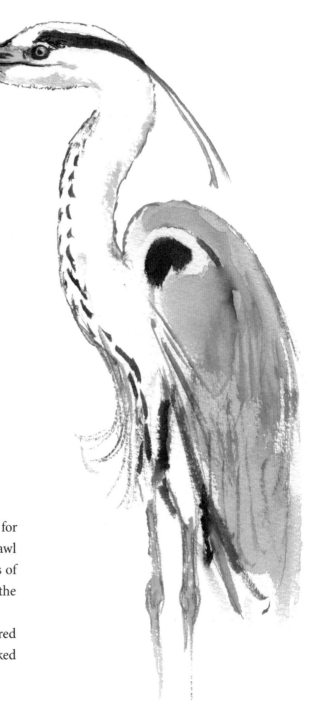

Grey Heron

Some people call the grey heron the 'granny fisher' for
it reminds them of an old granny with a grey shawl
out catching food for the grandchildren. In parts of
the west of Ireland, it is referred to as 'a sheila of the
bogs'.

In Elizabethan times, the grey heron was referred
to as a handsaw because the bird's bill and neck looked
like the saws used by carpenters.

Slow, heavy flight,
neck tucked in,
legs dangling

This bird could teach
us the art of embracing
stillness.

Dagger-like
beak

'When the wind is southerly, I know a hawk from a handsaw'

HAMLET, II, ii, WILLIAM SHAKESPEARE

Ballpoint pen

Sometimes when I'm outdoors
sketching I'm only armed with my
HB pencil, a ballpoint pen and
sketch pad. Later on I may work
them up into a pencil and ink
picture or a watercolour painting.

Ballpoint pen

Pencil

A grey heron seems to be able to
hold a perfect stillness as it
waits along riverbanks. What
might we learn? Stillness,
silence, patience.

46

Mute Swan

Mute swans are the largest and heaviest of our native birds. The male is known as a cob and the female as a pen. The young are called cygnets. This majestic bird may be seen in park ponds, rivers and lakes. It nests along riverbanks or among reeds and likes slow-moving, fresh or brackish waters. Couples create a strong bond and tend to mate for life.

Brush and ink

There is a
beautiful
side to life
that you
can discover
through the
creative act.

Kingfisher

This exotic-looking bird sits patiently on its favourite
perch, waiting for its favourite fish to swim by.

*The next time you are out
for a stroll along a river,
stream or lake, keep an eye
out for this jewel of the
waterways.*

Common Frog

We only have one species of frog in Ireland, which is common and widespread. They vary in colour from greenish-brown to buff olive with dark blotches. The male has swelling on its first finger to hold the female during mating season. The common frog is more agile than the toad, and hops and jumps to get around.

Nature can be a great healer — let it communicate with you. Soak in the beautiful atmosphere; let it calm and strengthen your spirit.

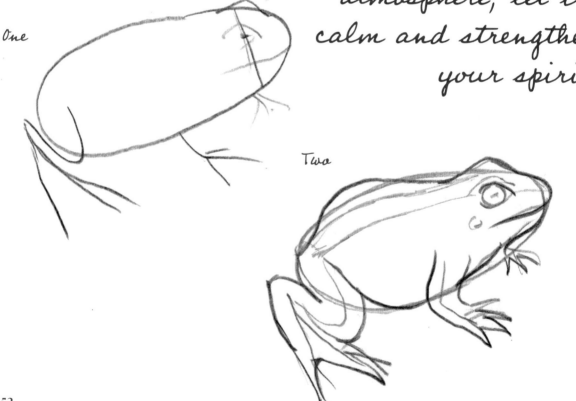

One

Two

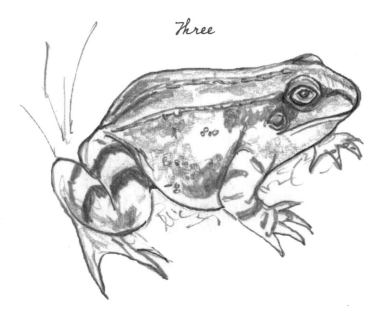

Three

Four

Natterjack Toad

Ireland's only species of toad, the natterjack, is very rare and only found in two protected locations in Kerry and Wexford. It tends to walk in a shuffling motion, but will jump when alarmed or feeling threatened.

Little Egret

A close relative of the grey heron, this elegant bird nearly became extinct in the late nineteenth century when the fashion industry used its exotic feathers to make hats and other accessories. Once a rare sight in Ireland, the little egret is now breeding successfully.

A glint of sunlight shows off the egret's dazzling white plumage as it moves along the water's edge.

A solitary
cormorant
perches near
the water's edge
as the gentle
autumn day
awakens.

Atlantic Salmon

Spending most of its life at sea, the Atlantic salmon makes an extraordinary journey home to breed, eventually following the river upstream to the very place it was born.

A cool, damp breeze off the sea
can awaken memories within.

An encounter with a wild salmon

When hunter-gatherers arrived in Ireland in the Mesolithic period, many settled along riverbanks. They developed a special reverence for the species of fish we know as the salmon. An important fish in their diet, it was tasty, and the large bones could be removed easily. More importantly, it was plentiful in midwinter when other food sources were scarce.

One mild November evening, my nephew Roger and I walked along the River Dodder. After a time we stopped at a waterfall near Milltown.

As we sat and chatted while listening to the sound of the rushing water, to our amazement, two large salmon leapt from the lower part of the river into the air to overcome the waterfall. One succeeded. One failed and landed right beside us. We hurried to pick it up; it was large and heavy. With it thrashing about in our hands, we returned it to the river. Roger and I were beaming with excitement and amazement at what had just occurred.

Dipper

The dipper has dark brown plumage and a white chest, a stout dark bill and black legs. Found along fast-flowing rivers and streams, it feeds on aquatic insects, small fish and underwater crustaceans. It usually builds its dome-shaped nest under old bridges or tree roots near water. Its flight is fast and low over the river or stream.

Dipper on a stream, Co. Wexford

Blackbird-like in shape and size but with a white bib, the dipper can claim the title of Ireland's only aquatic songbird.

One

Two

Three

Four

Dipper

Grey Wagtail

This slender, elegantly shaped bird is found along waterways. Although it gets its name because of the grey plumage on its head and back, it has a surprising amount of yellow on its underparts, especially during the breeding season.

I'm lucky to see this bird on a regular basis as it frequents a stream that flows near a public house in Wicklow town. It's only been seen drinking water there.

The rising sun begins
to sweep away the
mist over the lake,
blushing the water
with a dappled golden
light.

Crested lobe

The yellow iris,
also known as
the yellow flag

Outer petals
hang down

Fine veins
on leaf

Sharp leaves
may cut you
if not handled
carefully

Found in
shallow
waters

69

Mallard

The mallard is the most familiar of all of our ducks. Found throughout the countryside, it is mainly seen by rivers, lakes and on canals. Often it's one of the first wild creatures young people encounter when they are brought to a public park to feed the ducks. Mallards are the largest of the dabbling ducks. When I was growing up, I often visited Herbert Park, and the ducks in its pond were well fed by local lady Miss Green.

I remember walking by the pond years later, looking at the ducks and imagining that they still held the memory of Miss Green somewhere inside their little brains. The great English poet William Wordsworth said that poetry is emotion reflected in tranquillity. Here is a young man's poem for Miss Green, that gentle soul.

Encounter

You see her
Breaking bread by the water
In the early morning park
Neatly and tidily
She stoops
Scattering crumbs to the wind.
'Here duck duck duck'
She chants
As her feathered family
Hurtle forward
Shattering their mirrored path
Then all too soon
They sail away
Having gulped down her offerings
Leaving her by the water's edge.
Alone again
Carefully she folds her brown-worn bag
Returning home
To her family of memories.

Fluffy mallard ducklings can stand and swim within a day of hatching and follow their mother to water.

Snipe

The snipe's long, straight beak allows it to probe for molluscs and worms deep underground. During aerial courtship displays, the male snipe vibrates its tail feathers in mid-air, making a rhythmic drumming sound.

A wealth of life may be found
in the sparkling waters of
streams and brooks.

Reed bunting

OPEN SKIES AND SEA-WAVED SHORES

Like ancient mariners, the cormorants watch as the mist rolls in from the sea.

Cormorant

Although they spend most of their time in the sea and along rivers, cormorants do not have fully waterproof feathers, so they pass a great deal of the day drying off their bodies in a 'heraldic' posture.

The shore is a
symphony of
sound, sea
and sand.

Shag

The shag, smaller than the cormorant, shows off its
crown tuft during the breeding season. It is a common
sight along rocky shores and is rarely seen inland,
unlike the cormorant.

Curlew

The sound of the curlew has inspired many poets and nature writers with its haunting call that echoes the eerie sound of the estuaries and reminded people in bygone days of souls lost at sea.

Grey Seal

Many tales are told about these remarkable creatures, the 'people of the sea'. One such story tells of Noah's ark being closed when the deluge came. Through God's mercy, a few of the good people were turned into seals to survive the flood.

The gift of new life: a grey seal pup with its milky white fur.

Grey seal

The ragged cliffs are silent now
No hint of mighty colonies of razorbill, guillemot and kittiwake
That tussled for ledge-space in summer light
On dazzling, dizzy heights
Occasional ghostly forms of gannet and fulmar
Flit in and out of the silent cloaking fog
Moisture coats my binocular lenses
As I wait patiently on slippery rocks
Watching the weather closing in.

Paper buckles and hands become numb
As I clutch pencil and pad to capture elusive forms
I know they're there.
I hear Siren calls from shrouded island homes
Those eerie songs sing of 'Selkies'
Beautiful people of the sea
That come to visit our land-locked race
But never staying too long
Out of their seal-fur coats
Always ready to return
When the sea calls.

Darkness embraces the pearly fog
It's time to leave and try another day
But wait! A dark shape appears
Below the white foam
A grey seal breaks the surface skin
Staring at me with wild curiosity
A lightening sketch of lines and shapes
Moments later it slips away under the marbled water
I've captured its form
In those dark liquid eyes
Has it captured mine?

Puffin

During the breeding season, puffins will search out
a rabbit's hole on the sloping, grassy clifftops to raise
their one and only chick.

Observing nature can help you wake up to yourself.

Harbour Porpoise

The waters of Ireland are now a protective sanctuary for whales, dolphins and porpoises, known as cetaceans. The porpoise, unlike the dolphin, does not have a long beak, and it is very wary of contact with humans.

Some 24 species of whales and dolphins have been recorded in Irish coastal waters.

Bottlenose Dolphin

Of all the dolphin species found in Irish waters, the
bottlenose is the one most frequently seen. Dolphins
hunt by using echolocation to find their prey. The
males are larger and longer than the females.

Nature has
its own secret
language.

Otter

The otter, also known as the water dog, is found anywhere there is water: rivers, lakes, streams, estuaries and coastal areas. Listen out for its whistling call if you happen to be walking along an estuary at dusk.

In search of otters with more than a bite in the wind

Looking out at the spectacular coastal scenery that was being wracked by fierce Atlantic waves, I was on a mission to see otters. This was by no means guaranteed; otters are famously secretive and can be hard to find. A friend had shared his local knowledge with me, letting me know where he had seen a family of otters on a regular basis in a sheltered inlet.

I was grateful that the day was dry despite the biting wind that blew in from the sea. Carefully, I made my way to the sheltered bay where the sea was much calmer. With a pair of binoculars around my neck, a sketchbook and pencil in my jacket and a bar of chocolate in my pocket I was ready for the vigil.

Near a small sloping pathway, which may even have been made by otters, that led to a shingle beach I noticed their calling card, a 'spraint', on a small rock – in other words their droppings, a good sign that otters were about.

Soon I was rewarded by the sight of an otter in the swell. It dipped below the surface. I knew it would pop up soon. Although otters have a thick fur coat, they do not have a thick layer of blubber like seals to keep warm. Quickly it surfaced. It seemed to have a crab in its grasp.

What I did not realise was that there was a 'holt' nearby; that's a hole where otters live. Nor did I realise that there were young otter pups bounding towards me as I lay face down on the grass between two boulders. They did not pick up my scent because of the winds that circled about. One even jumped onto my back then away again. The other stopped, looked at me, nipped my thigh with its needle-sharp teeth, then hopped away.

With the mother still out at sea I quickly managed to do some doodles before departing as the evening began to close in.

We see a lot of amazing wildlife on television, but there's nothing more thrilling than actually seeing it in the wild. I was nine years old all over again.

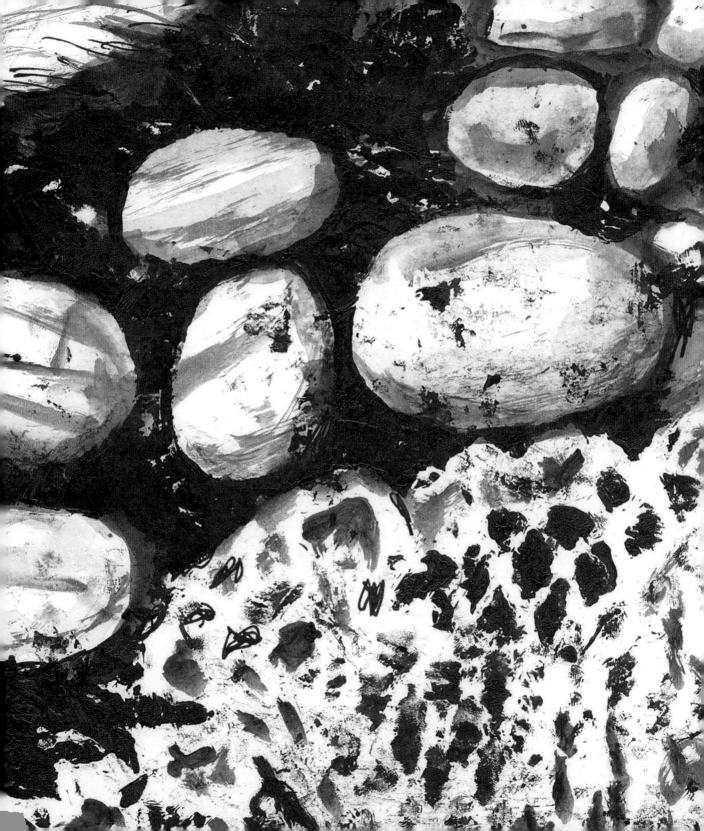

Chough

In a rush of wings, the chough takes to the air showing off its finger-like feathers, calling with a few clear notes.

The chough is the most attractive of the crow family, with its bright-red beak and legs.

Kittiwake

Kittiwakes form their nests on steep cliffs, protecting their young from predators. They are considered the most beautiful of the complicated and sometimes misunderstood gull family.

A lament for a dead kittiwake

Stone cold it lay
Upon the sea-wave shores
No one missed it from the sky
No winds had said it had fled

Where once it sailed
Through morning blue skies
And glided in the evening air
A silhouette against the radiant light

I once thought it possessed eternal youth
As I watched it in majestic flight
Alas no longer can I carry
Such hopes and idle dreams

Still I wonder
Can such a creature reincarnate again
Or have all its hours flown?

The sand has shrouded his body
Creating his sepulchre by the sea
Yet his eyes reveal a deep silence
An utter peace

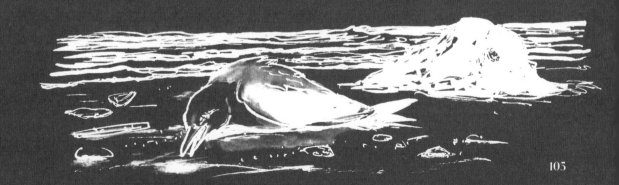

HIGHLANDS AND LOWLANDS

Merlin

This energetic hunter is mostly silent, except around its ground nesting sites on mountains, moorlands and blanket bogs. The male's plumage is blue-grey, and the female has browner colouring.

Red Grouse

The red grouse is rarely seen in flight, but it will sometimes take off explosively before settling into a glide. Though rare, it is widespread and can be found on mountains, moorlands and blanket bogs, where it mainly eats heather.

Male has a 'comb' over the eyes

Reddish chestnut colour

Legs covered with white feathers

Hen nesting

Common Buzzard

A daylight hunter, the buzzard was almost extinct in
Ireland by the early twentieth century. Now, however,
it has returned to our skies and can be seen scavenging
and searching for small live prey.

*After soaring high
in the sky on its
broad wings, the
buzzard rests
briefly in the
branches of a bare
tree.*

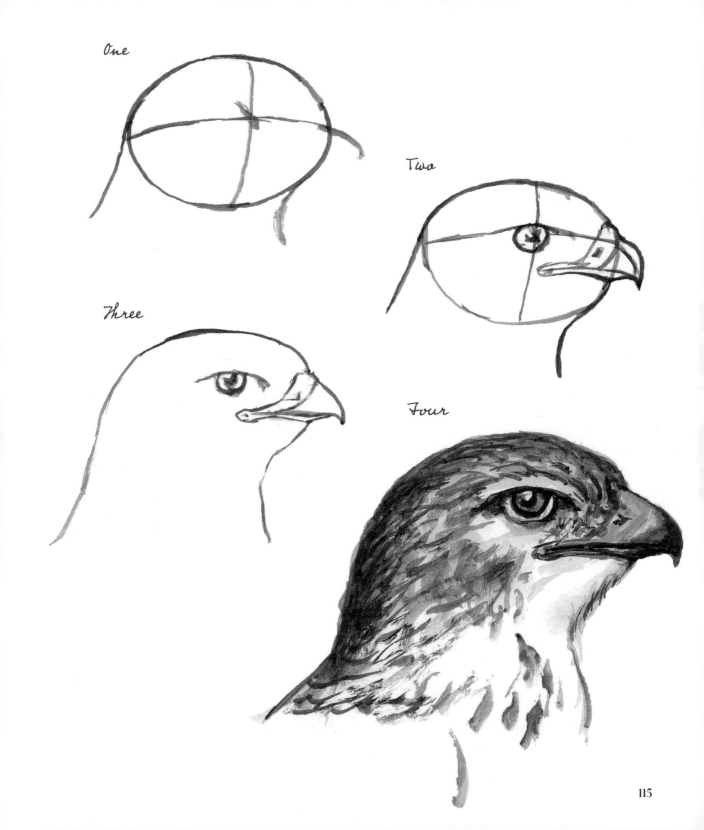

One

Two

Three

Four

115

Peregrine Falcon

This magnificent, deadly raptor nests on cliff faces and mountain ledges. Its flight is swift and strong, and it preys on other slower birds, such as pigeons, thrushes and gulls.

The natural world is full of mystery and beauty, but it does not easily reveal its secrets to the undiscerning eye.

An unfamiliar call pierces the air as I wind my way along a narrow pathway between the cliffs and the sea. Soon I am treated to the sight of a female peregrine, the fastest bird alive.

In search of the fastest bird alive

High across a cobalt-blue sky, one of the greatest aerial hunters made its appearance as it set out to hunt. This long-winged falcon meant business. From a high vantage point in the sky it targeted a flock of racing pigeons, unaware of it as they moved across the craggy mountain cliff slopes. This deadly assassin eyed and selected its quarry.

The falcon wheeled, tilted and arched its body. It already had a victim in its sights. With wings tucked back it dived, making a stooping plunge through the air for the kill. Suddenly a pigeon was knocked out of the air by the powerful talons and tumbled to the valley below. The panicked flock scattered in all directions. The peregrine soared back into the air triumphantly and then leisurely descended to claim its meal, taking it away to a favourite plucking post. I marvelled and at the same time was shocked at this wild, savage drama that had unfolded before my very eyes.

Weeks later I was asked to help rescue a large bird trapped in an industrial building in the city. It turned out to be a female peregrine falcon that may have chased feral pigeons inside the old factory. After a great deal of effort we managed to rescue the bird. Fortunately I have falconer gloves to deal with those sharp talons. The bird had jesses on its legs, so I knew it must have belonged to a falconer. I also knew the falconer would be most anxious to get his prize bird back again. It took me several days to locate the owner of this amazing specimen. During this time I had the opportunity to bond with the creature, to feed it with scraps of meat that I got from the local butcher. I remember holding it close, looking into its fierce wild eyes, sitting it on a perch to sketch it. It was a wonderful experience. What I found, despite its strength and wild nature, was that it had a calmness and allowed me to stroke it.

When I finally returned the bird to its owner he was overjoyed. In gratitude, he presented me with a book on falconry. I was just so grateful to have had the time with this 'Lord of the Skies' and, of course, to get to sketch it.

Lapwing

In the early morning light of dawn, the plaintive melancholic call of a male lapwing announces its arrival. In the damp stubble field, sitting on a clutch of olive-green eggs, the female is motionless and alert as she watches the male tumble from the sky calling out 'pee-wit, pee-wit', soon to join her.

The lapwing is a large wader with a distinct wispy crest that is longer on the male than the female. No other bird in Ireland or Britain has this plumage feature.

Wispy crest

Hare

The hares leap and zigzag about in open fields. It's the breeding season. Jack is looking for a Jill, chasing and frolicking about, and there's plenty of competition. It's the female who will box and kick, as she tests the strength of her suitors. After several bouts she will choose a partner. There is method in her madness!

Real wealth is to be enriched
by life, by true friendship, by
true companionship and, if
you're lucky enough, by true
love.

Rabbit

Rabbits live below ground in burrows to hide from predators. Smaller than the hare and more stocky, with a white tail, the young are called 'kittens'.

Corncrake

This elusive bird, so difficult to detect in a meadow, has a trick up its sleeve. Its call has a ventriloquial quality, making it very difficult to pinpoint where it's coming from.

I'm sitting in a thatched cottage on the island of Inishbofin. There is a warm glow from the turf fire. The air is filled with the pleasant smell of the peat. It's a soft evening outside, as the locals would remark. Close to the cottage is a field carpeted with a variety of wildflowers and long sweet-smelling grasses. In the dense cover I can hear the repeated call: 'crex-crex'. It's a corncrake, a secretive bird, seldom seen. But its far-carrying rasping call is music to my ears.

From a high
vantage point
the golden eagle
surveys her
domain.

Fly eagle fly

Like silence you came from your castle of rocks
Riding the north wind down to the sea
Soaring and gliding on proud stiff wings
Past ancient cliffs
That held your memory, locked in
Her sacred stones

On currents of warm air you sail
Tracing hidden pathways through the sky
Do your wings carry secrets
From mountain to mountain and from
Rivers to the sea?

The earth coloured your body
While the sun burnished your head
Your wild screams tear the silence
As you scan with that fierce stare
Across landscapes that earth-bound spirits share

Do not listen to their plea
Or be charmed by their weasel words
They who kill the visions
And dream only empty dreams
They can offer only chains and
Would clip your precious wings
So fly eagle fly on wings, so free!

Skylark

High above in a blue sky over a damp buttercup meadow hangs a skylark on fluttering wings. There it sings its sweet liquid melody.

You must be enthusiastic about life.

Swallow

An agile flier, the swallow may be seen darting, zooming and flitting with mouth agape, catching flying insects. Inclement weather drives swallows into a state of torpor. This lethargy is one step above hibernation. They can remain in this state for several days until the weather improves.

The site-faithful swallow makes a long journey back to the old barn to rebuild the saucer-shaped nest and raise a new family.

The male swallow collects material for the nest, which is made of mud pellets, and the female builds it.

Rook

High up in tall trees close to farmlands, these social birds build their rookeries, with large flocks living together. Noisily they will head to a newly ploughed field in search of leatherjackets and wine worms, their favourite food.

What might we learn from the kestrel? To get an overview of things, a bird's-eye view, and to see things from a different perspective.

Kestrel

Over the harvest field the kestrel will hover, spreading out its fan-like tail. Its body stays almost motionless, apart from the quivering wings. Big, bright eyes scan the field for a rodent or large insect. It drops like a stone from the sky to catch its prey.

Red Kite

What a welcome return to our Irish skies is the red kite. This magnificent raptor, unlike the common buzzard, shows off a forked tail.

Where attention goes, energy flows.

Pheasant

The cock pheasant with its superb colourful plumage struts about at the edge of the long grasses. The hen often sits secluded nearby in some bramble bushes and herb Robert, her plumage blending in with the earth colours.

Barn Owl

With its ghostly white form flying silently over an old cemetery, giving out the occasional loud, blood-curdling scream, this bird is probably responsible for more tales of terror than Bram Stoker and Sheridan Le Fanu put together.

The barn owl is a farmer's friend, helping to keep down the rodent population. All it needs is a roof over its head to roost and nest in.

After a night-long vigil I am rewarded with the ghostly sight of a barn owl as it leaves its home in the old castle ruin.

I've had the chance to take care of injured owls over the years. They are remarkable birds. It's always a joy to draw and paint them. I've even featured them in some of my children's novels.

DON CONROY

Ghost of the night

As twilight falls upon the ruin
A blackbird scolds departing light
While woodcock rode at the moon
And bats take flight
Those messengers of night.

A meditative silence around me crept
A silence so intense
Had the world stopped to rest?

Alone I waited
With excitement and care.
For beware the castle lore
Of the demons and the ghosts
That stalk the night once more.

From out of dwellings dark
Came a hissing and a snoring. Then a shriek!
My heart did leap.
No ghost or shade I saw
On silent wingèd flight
But a beautiful white bird of night
Who made his home
Where king or squire once did rest.

At dusk, on quiet
wings, the barn
owl can be seen
quartering the fields
in search of its
next meal.

ON YOUR DOORSTEP

Blackbird

This garden visitor loves to sing from a high vantage point. The female can sometimes be mistaken for a song thrush, with her speckled brown feathers. The male, however, is conspicuous, with black plumage and a bright-yellow bill.

Fine greenish-blue eggs with reddish markings

Blackbird nest

European Robin

The robin is a frequent and much-loved visitor to urban and rural gardens. Fiercely territorial, it is the only bird in Ireland that keeps singing through the winter.

Robin egg (20mm)

Listen to the grand symphony of the dawn chorus in springtime.

A young European robin, all speckled. It will have to wait for autumn to get its adult red breast.

DON CONROY

Hedgehog

A hedgehog sniffs and shuffles through a thick layer
of fallen leaves under the trees.

In the springtime there
seems to be a return to
life. Everything appears
newly made again.

One

Two

Three

Four

Sparrowhawk

Pied Wagtail

A regular and welcome visitor at our kitchen door is the pied wagtail, always looking for breakfast scraps. Affectionately known as the willy wagtail, this elegant bird has black-and-white plumage and a long bobbing tail.

The male and female may look similar, but on closer inspection you will discover that the male's colouring is much darker. It is commonly seen briskly walking along kerbs or footpaths in parks or along the pier at a harbour. Despite its showy appearance, it is often overlooked by passers-by.

It's a wondrous world –
just look!

yellow spearwort

House Sparrow

The social house sparrow will often take a dust bath with its fellows to smother parasites in the feathers and absorb excess oil. Often living very close to humans, this bird nests in man-made structures, such as gutters, street lamps and birdhouses.

We're all on this journey together, so value life, value each other and look for the great creativity that is lying dormant within you, waiting to be awakened.

One

Two

Three

Four

Goldfinch

In winter, the goldfinch extracts seeds from the dead seedhead of the teazel herb.

165

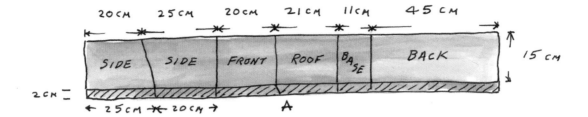

Nest boxes can be very useful. Two simple designs will do for starters, for smaller birds.

The tit family prefer a closed nest box with an entry hole. The size of the hole is important: 25mm for blue tits, 28mm for great tits.

Robins will use open-fronted boxes. Use strong timber and treat it with a harmless wood preservative on the outside to prolong the life of the box.

These birds will build a nest inside the box. Throw out the old nest and clean when the season is over.

You can modify this design to make a bat box with an entry slot at the bottom. Make sure to keep the slot narrow, between 15 and 20mm.

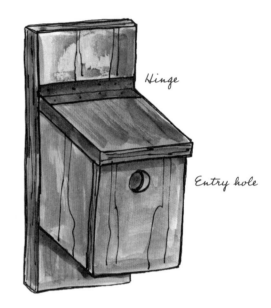

Bat box

Narrow entry slot

Open-fronted box

One can see the
beauty of nature
and, sometimes,
its fragility.

Well-stocked bird feeders are a welcome sight to hungry garden birds in winter-time.

Chaffinch

Long-tailed tits

Chaffinch

Starlings

Song thrush

Song thrush nest

Song thrush egg
(27mm)

Blue tit

Great tit

Woodpigeon tail
feather

Wren

Dunnock

Great tit

Goldcrest

171

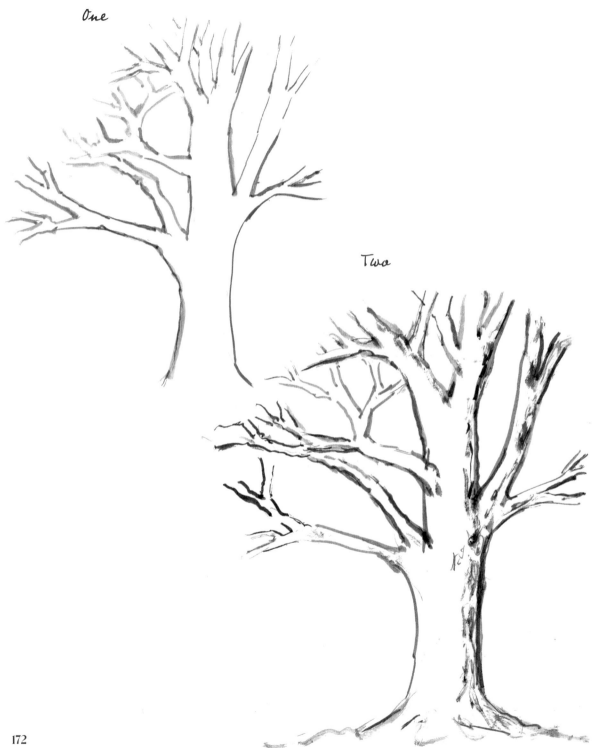

One

Two

Three

Swifts

Song of the wren

Proudly you appear on top of a bramble bush
A prominent position
It's your singing perch
Feet locked and tail cocked
You're ready to shine
To serenade your beloved

Normally shy and elusive
Shunning the limelight
Skulking among hedges
Your chocolate-brown plumage
Blending in amongst the undergrowth

Sometimes you 'whirr' about
Searching for insects with keen eyes
And tiny sharp pointed bill

The season is bright
You've built several cock nests
The hen has chosen one
To lay her clutch
Of tiny glossy white eggs

You ruffle your feathers
On your short plump body
It begins – your aria
A little bird with a big voice – a pocket Caruso

Close by she listens
Comforted by your trills
While she broods the future songsters.

'Pale primroses,
that die
unmarried'

THE WINTER'S TALE, IV, iv, WILLIAM SHAKESPEARE

Waterfall of wildflowers

Spring's symphony of colourful wildflowers begins slowly, then it spreads to blanket the countryside. Of the many beautiful wildflowers, the primrose is one of my favourites. A delicate five-petalled yellow flower with heavy green-veined leaves, its Latin name is *primula vulgaris*. '*Primula*' means first rose. Mainly a plant of woods and hedgerow banks, the primrose is said to herald springtime. The first flower starts singly on a shaggy wiry stalk, followed by several long-stalked flowers growing from the rosette. The second Latin word, '*vulgaris*', means common, although it's far less common nowadays than it used to be. Primroses rely on insects to help them pollinate; if there are few insects around, they may not get pollinated.

I've been very privileged over the years to encounter and sometimes befriend some remarkable people. One feels enriched by the experience. One such person was an American gentleman, Thomas Berry. How to describe him? A Catholic monk, a writer, a theologian, a cosmologist, an earth scholar and a storyteller. We were invited to give a talk relating to the environment. My talk was on how nature can inspire us in a creative way. His talk was looking at the environment from a spiritual and holistic perspective, reminding the listeners that 'we are not a collection of objects on the planet, but a communion of subjects that we need to discover humanely through nature – physical and of the spirit.' He was suggesting that we look at the world like a child and marvel at its wonders, which are daily being revealed to us.

After our lectures and lunch we had the opportunity to spend time together over a coffee. As we sat there I commented on a framed poster of wildflowers on the wall behind us. 'You know, Tom, one of my favourite wildflowers is the primrose.'

Suddenly, he really looked at me, then a moment later a tear escaped from his eye.

'Don, one of my most treasured memories is from when I was six years old. I was strolling with my mother in a local woods. I was holding her hand, I can still feel the softness of her skin. As we left the woods through a small clearing my mother exclaimed with excitement, "Look, honey!" – there was a yellow waterfall on a bank near a creek. I could see small yellow flowers spilling down the bank. "They're primroses, honey, the first rose." Because my mother was excited at seeing them, I became excited too.'

He took a sip of his coffee and sighed.

'That woods is long gone now, and the creek of course, so is my blessed mother, Lord rest her soul.' Then with a gentle smile he said, 'You know, Don, when I get stressed I simply close my eyes, meditate on that day, and I can conjure up that yellow waterfall, see all those lovely primroses spilling down the grassy bank.'

Acknowledgements

With thanks to very many people for their friendship and support over the years, in particular: Dr Simon Berrow; Juanita Brown; Richard Collins; Cliona Connolly; Dave Daly; Gordon D'Arcy; Ray D'Arcy; Eric Dempsey; Ian Dempsey; Leo and Clare Hallissey; Tom Hayden; Nuala Holloway; Gabriel King; Dermod Lynskey; Alan McGuire, South East Radio; *Mooney Goes Wild*; Derek Mooney; Killian Mullarney; Richard Nairn; Éanna Ní Lamhna; Jim O'Connor; Sean Olohan; Fintan O'Toole; Lorcan O'Toole; Don Scott; Nicola Sedgwick; Gordon Snell; Shea Tomkins, *Ireland's Own*; and Ann Wilson.

Thanks as well to: Birdwatch Ireland; Irish Wildlife Trust; Irish Peatland Conservation Council; Irish Raptor Study Group; Irish Whale and Dolphin Group; National Gallery of Ireland; Natural History Museum of Ireland; Royal Society for the Protection of Birds (RSPB); Wexford Wildfowl Reserve; and Wildfowl and Wetlands Trust (WWT).

To all the wonderful teachers and librarians I have had the pleasure of meeting and working with throughout the years, thank you.

Finally, a big thank you to all at Gill Books for their excellent publication, and their friendship.